WILD CATS

TIGERS

by Alissa Thielges

AMICUS | AMICUS INK

stripes

whiskers

Look for these words and pictures as you read.

tail

cubs

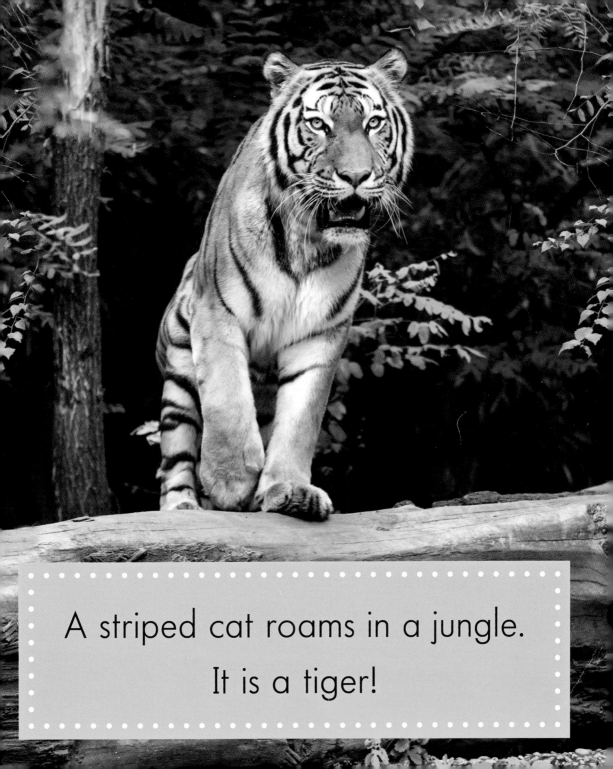

A striped cat roams in a jungle.
It is a tiger!

Tigers are the biggest wild cat.
They live in Asia.

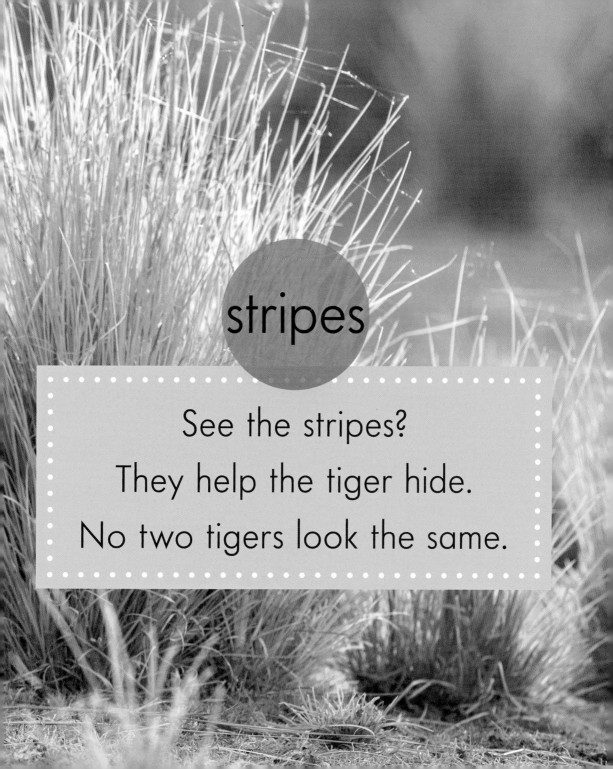

stripes

See the stripes?

They help the tiger hide.

No two tigers look the same.

See the long tail?
It is used for balance.
A tiger makes quick turns.

tail

See the whiskers?
They feel for things.
They help a tiger
hunt at night.

whiskers

See the cubs?
They stay with mom for two years.
They learn to hunt.

cubs

Tigers love water.
They swim and play.
Splash!

stripes

whiskers

Did you find?

tail

cubs

Spot is published by Amicus and Amicus Ink
P.O. Box 1329, Mankato, MN 56002
www.amicuspublishing.us

Copyright © 2021 Amicus.
International copyright reserved in all countries.
No part of this book may be reproduced in any form without written permission from the publisher.

Library of Congress Cataloging-in-Publication Data
Names: Thielges, Alissa, 1995- author.
Title: Tigers / by Alissa Thielges.
Description: Mankato, MN : Amicus/Amicus Ink, [2021]. | Series: Spot | Audience: Ages 4-7 | Audience: Grades K-1
Identifiers: LCCN 2019036108 (print) | LCCN 2019036109 (ebook) | ISBN 9781681519326 (library binding) | ISBN 9781681525792 (paperback) | ISBN 9781645490173 (pdf)
Subjects: LCSH: Tiger--Juvenile literature.
Classification: LCC QL737.C23 T473536 2021 (print) | LCC QL737.C23 (ebook) | DDC 599.756--dc23
LC record available at https://lccn.loc.gov/2019036108
LC ebook record available at https://lccn.loc.gov/2019036109

Printed in the United States of America

HC 10 9 8 7 6 5 4 3 2 1
PB 10 9 8 7 6 5 4 3 2 1

Gillia Olson, editor
Deb Miner, series designer
Ciara Beitlich, book designer
  & photo researcher

Photos by Alamy/Mark Passmore 12-13; Getty/John Conrad 14-15; Getty/dikkyoesin1 cover, 16; iStock/GlobalP 1, 10-11; iStock/Photocech 4-5, 6-7; Minden/Edwin Giesbers 8-9; Shutterstock/Ewa Studio 3